Where Did The Sun Go?

The Great American Total Solar Eclipse

By: Misty Carty, Ph.D.

On a special day in August, Mommy woke me up before the Sun had risen. She said it was the day of the *total solar eclipse* – the day the Moon would briefly cover up the Sun!

Sunrise on August 21, 2017 — the day of the Great American Total Solar Eclipse.

My family dressed and ate breakfast quickly. Then, we drove to Casper, WY – a city along the *path of totality*. Mommy said we have to be within this path to see the Moon cover the Sun.

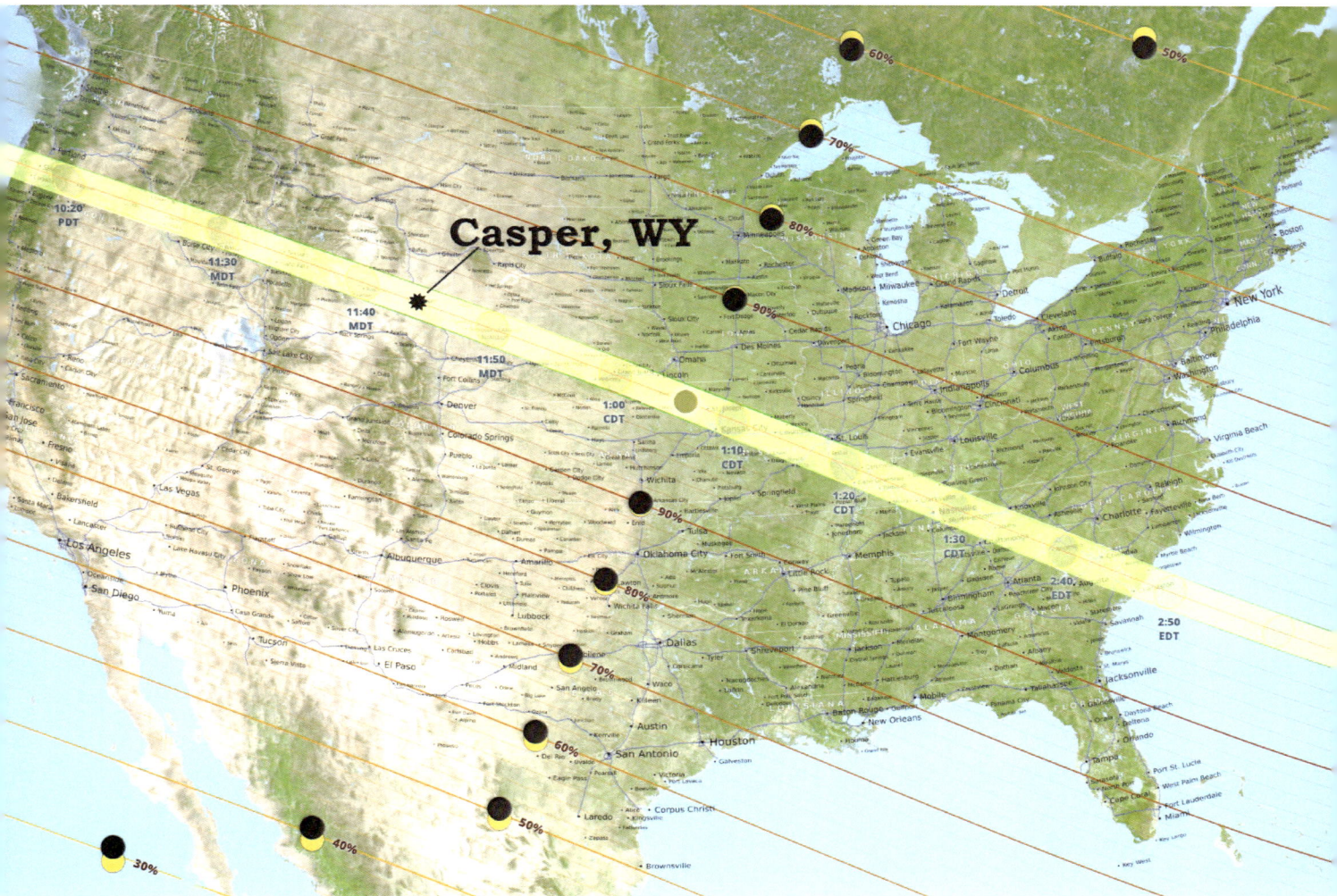

The yellow line shows the path of the total solar eclipse across the United States - where the Moon's shadow will fall on the Earth!

There were lots of cars full of people when we arrived in Casper. We parked and found a grassy spot to sit and view the eclipse.

Then, Mommy gave me special glasses. She told me that before the Moon covers the Sun completely, it will only be blocking part of it. I must wear special *eclipse glasses* to look at the Sun and see the *partial solar eclipse*.

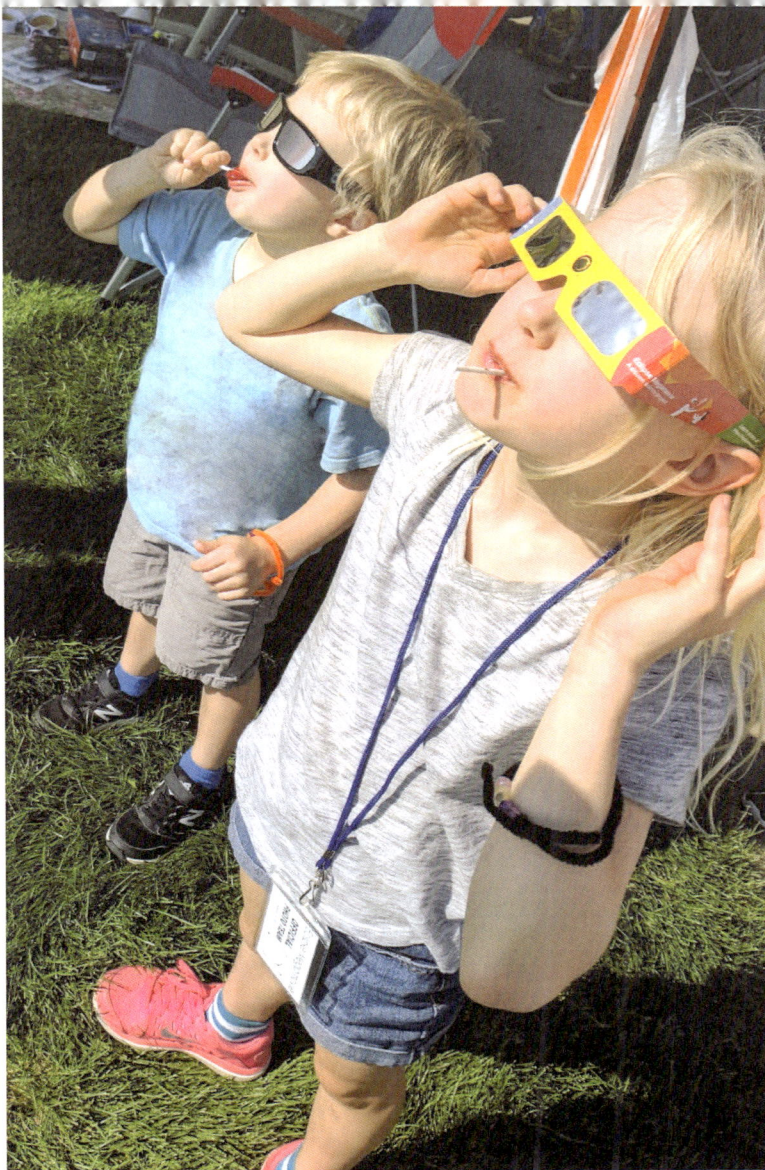

We can still see the Sun's light during a partial solar eclipse, which can damage our eyes quickly if looked at without proper eclipse glasses or filters.

At 10:22 a.m., the partial eclipse began with the Moon sliding in front of the top right corner of the Sun. I put on my eclipse glasses like Mommy told me to and looked up. The Sun looked like a cookie with a nibble taken out of it!

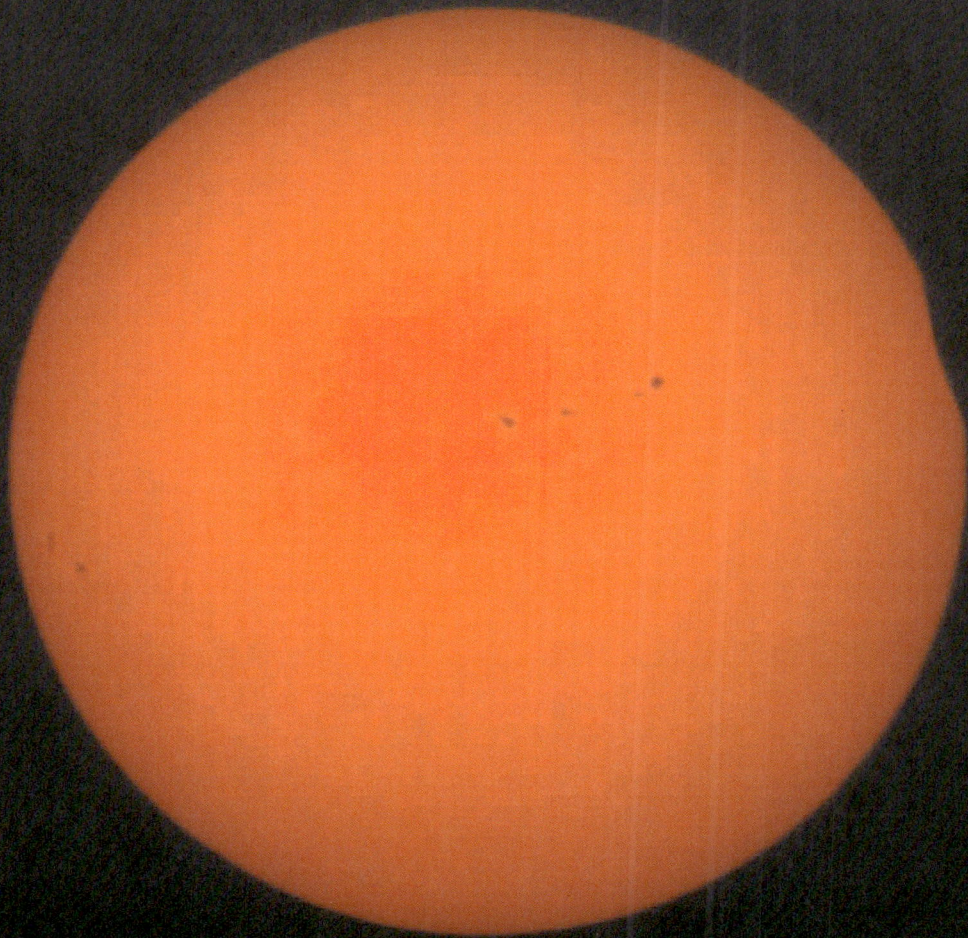

The start of the solar eclipse. The dark arc on the right of the Sun is the Moon beginning to pass in front of it. The small dark spots on the Sun are sunspots!

Over the next hour, the Moon moved more and more in front of the Sun.

10:23 a.m.

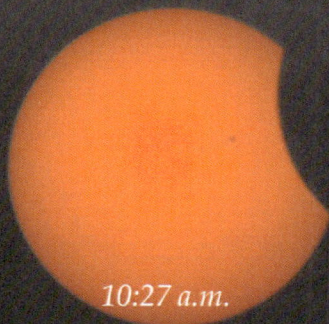
10:27 a.m.

10:34 a.m.

10:43 a.m.

10:47 a.m.

10:53 a.m.

10:59 a.m.

11:05 a.m.

11:10 a.m.

11:16 a.m.

11:21 a.m.

11:28 a.m.

In addition to eclipse glasses, Mommy and I used pinhole viewers to see the partial eclipse. We also made a simulated eclipse with a model of the Earth and Moon.

Using a pinhole viewer to see the partial solar eclipse.

Earth   Moon

A scale model of the Earth (1" across), Moon (1/4" across), and the distance between them (30"). Top right corner: zoomed in view of scale Moon.

A simulated eclipse using the scale model of the Earth and Moon. The dark circle is the Moon's shadow falling onto the Earth — a scale total solar eclipse.

At first, I couldn't tell the partial eclipse was happening without my eclipse glasses or a pinhole viewer.

But as the Moon covered more of the Sun, making it a crescent shape, I noticed the sunlight seemed strange. It wasn't nearly as bright out as when the partial eclipse began and my shadow looked different - some of its edges were sharp while others were spread out.

The shadows of the fingers in the bottom hand have sharp edges. The edges of the shadows in the top hand are less defined and more spread out.

Using my eclipse glasses, I watched as the Moon covered up the last bit of the Sun.

Then, there was no more Sun. "Where did the Sun go?!" I exclaimed.

11:33 a.m.

11:37 a.m.

11:41 a.m.

Mommy said it was now safe for me to take my eclipse glasses off. As I did, the last bit of sunlight flashed from behind the Moon, making the pair briefly look like a diamond ring.

Just before totality, the last of the Sun's light, passing around mountains and through valleys along the Moon's edge, gives the appearance of a diamond ring.

At 11:42 a.m., darkness swept over the hill and the sky became a deep, dark blue just like at twilight. We were standing in the Moon's shadow!

Looking up, I saw the Sun completely covered by the Moon! It had a glowing ring around it – the Sun's atmosphere, the *corona*.

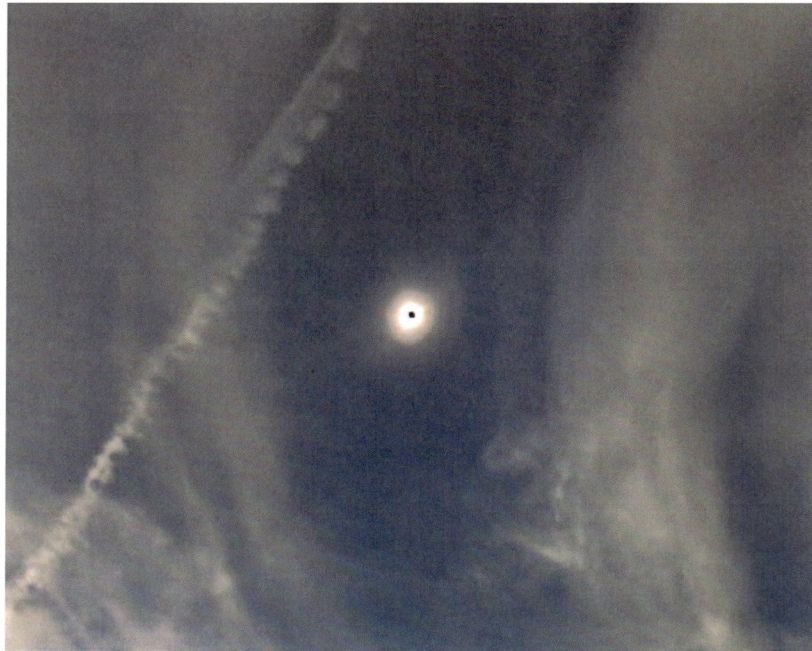

Wide angle view of the total solar eclipse.

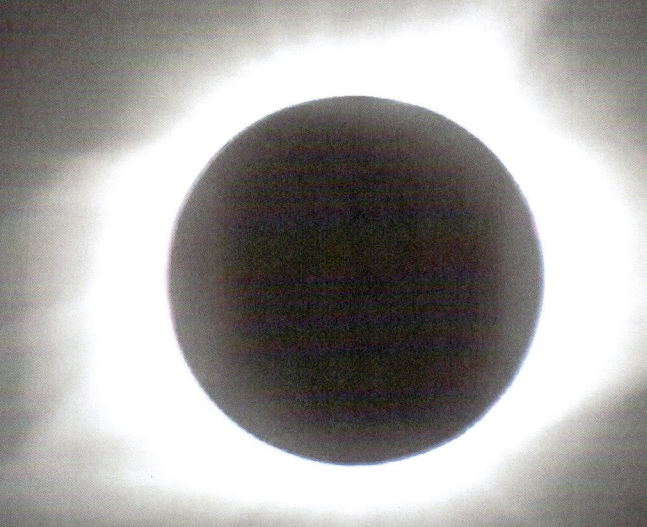

Zoomed in view of the total solar eclipse. The dark circle at the center is the Moon completely covering the Sun. The glow around it is the Sun's corona.

The sky was dark enough I could see a few stars and the planet Venus! And, the horizon was a circle of sunsets painted pink and orange.

Venus became visible during the total solar eclipse!

During the total solar eclipse, the sky was a dark blue with a beautiful sunset on every horizon.

The Moon stayed in front of the Sun for almost two and a half minutes!

As it began to slide off, the sky brightened first behind me. Then, the darkness quickly receded across the sky.

The sky becoming lighter as the total solar eclipse ends.

After the total eclipse, the sky was no longer dark but the sunlight was dim, like before *totality*, casting strange shadows.

Many people began packing up to leave. Mommy asked what I would like to do. I wanted to stay till the very end! Using my eclipse glasses, I watched as more Sun appeared.

11:52 a.m.

11:58 a.m.

12:05 p.m.

12:11 p.m.

12:16 p.m.

12:29 p.m.

12:37 p.m.

12:42 p.m.

12:50 p.m.

12:53 p.m.

12:58 p.m.

1:03 p.m.

As the partial eclipse was coming to an end, I saw only a tiny bit of the Moon covering the bottom left corner of the Sun. It again looked like a cookie with a tiny bite taken out of it!

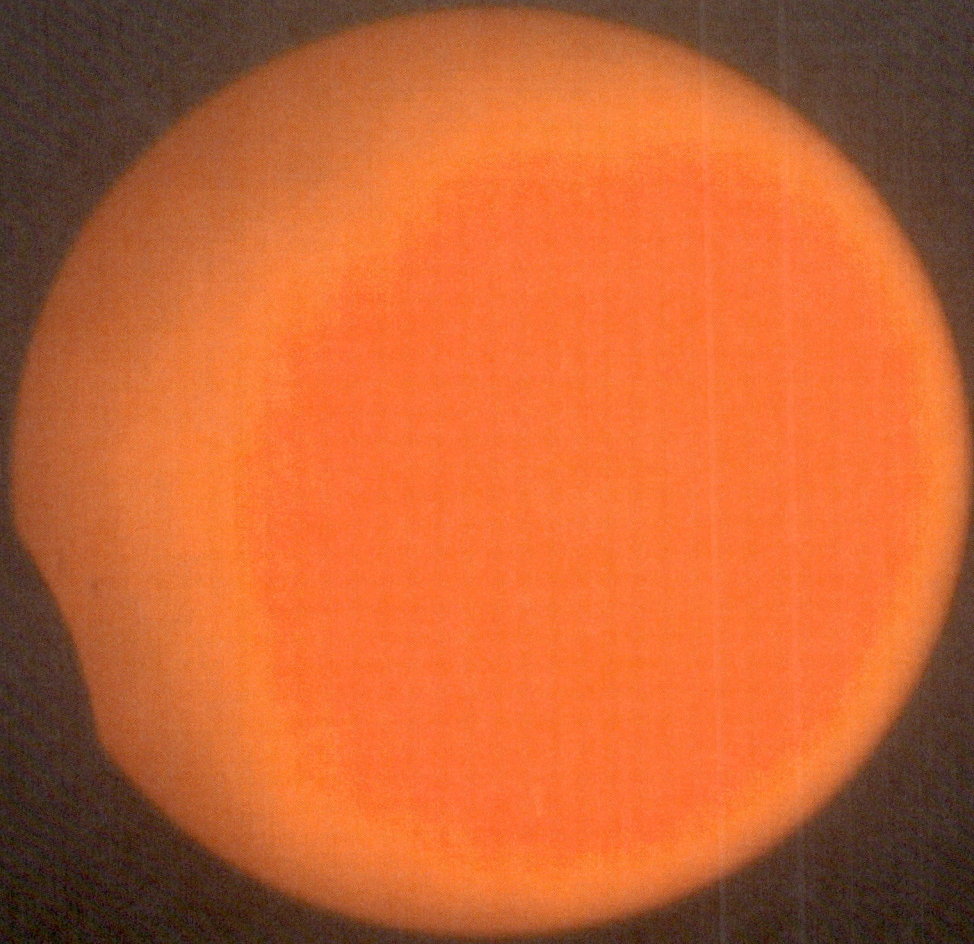

The end of the solar eclipse. The dark arc on the left of the Sun is the last of the Moon passing in front of the Sun. The small dark spots on the Sun are sunspots!

At 1:09 p.m., the Moon no longer covered the Sun. I was tired and excited at the same time. With the entire solar eclipse over, it was time to head home.

The Great American Total Solar Eclipse completed!

As we drove, I thought about all the amazing sights I had seen.

And knew, one day, I would travel to witness a total solar eclipse again!

Glossary

Corona: The outer most layer of the Sun's atmosphere. Astronomers can study the corona when the sunlight from the surface of the Sun is covered, like during a total solar eclipse.

Eclipse glasses: Eclipse glasses are essential to protect your eyes when viewing the Sun, even during a partial eclipse. They block 99.999% of visible light from the Sun and 100% of harmful ultraviolet and infrared light.

Partial solar eclipse: The Moon covers part, but not all of the Sun. Partial solar eclipses occur before and after a total solar eclipse. But they can also occur on their own.

Path of totality: The area on the Earth a person can see a total solar eclipse.

Total solar eclipse: The Moon covers the visible surface of the Sun completely.

Totality: The period when an eclipse is total.

For Parents and Educators:

Introducing science to young children is fun! As any parent knows, kids are little scientists; experimenting and interacting with the world around them every day. Already curious, children love furthering their vocabulary and knowledge about the objects and actions they experience.

In this book, *Where Did The Sun Go?*, your child will get to experience the Great American Total Solar Eclipse! The eclipse occurred on August 21, 2017, the first to cross the United States in almost 100 years. People across the country were able to witness at least a partial solar eclipse. Those who traveled to the path of totality - or lucky enough to live within it – were treated to a wondrous total solar eclipse. With this book, your child will observe all the phases of the Great American Total Solar Eclipse and get to experience what it was like to be in the path of totality!

As you read this book with your child, please use the following information to deepen your science experience.

Photograph page 5 © NASA, visualizations by Ernie Wright, released on December 13, 2016. You can explore this eclipse map and others at https://svs.gsfc.nasa.gov/4518.

About Total Solar Eclipses

Total solar eclipses occur when the Moon passes between the Sun and Earth - when the Moon is at its new phase. There are between 2 and 4 solar eclipses a year. But their narrow path and the fact that the Earth is covered mostly by water, make many of them difficult to witness.

Solar eclipses do not occur every month when the Moon is new because the Moon's orbit around the Earth is tilted slightly (by ~5°). Most of time the Moon passes either above or below the Sun at its new phase. However, as the Moon circles the Earth its inclined orbit shifts a little – it precesses – causing the occasional perfect alignment of the Sun, Moon, and Earth.

Another key factor for a total solar eclipse to occur is the distance of the Moon from the Earth. The Moon's orbit is not perfectly circular. This means that the Moon is not always the same distance away from the Earth: it has a closest point – perigee, and a furthest point – apogee. When the Moon is at perigee, its size on our sky appears slightly larger than the size of the Sun! Therefore, when the Sun, Moon, and Earth align and the Moon is at its closest to the Earth, we will see the Moon completely cover the Sun. However, when the Moon is at its furthest from the Earth, its size on our sky appears slightly smaller than the size of the Sun. If the Moon is at apogee when the Sun, Moon, and Earth align, we see what is called an annular eclipse or a ring of fire – the Moon does not completely cover the Sun!

www.ingramcontent.com/pod-product-compliance
Lightning Source LLC
Chambersburg PA
CBRC101143030426
42336CB00007B/71